State of Rhode Island, &c.

MEMORIAL OF PROF. RIDGWAY,

GEOLOGIST AND MINING ENGINEER,

IN RELATION TO THE

Coal Field of Rhode Island,

PRESENTED TO THE

GENERAL ASSEMBLY,

AT ITS

JANUARY SESSION, A. D. 1868, AND PRINTED.

RE-PRINTED BY ORDER OF THE GENERAL ASSEMBLY AT ITS JANUARY SESSION, 1870.

PROVIDENCE:
PROVIDENCE PRESS COMPANY, PRINTERS TO THE STATE.
1870.

MEMORIAL.

To the Honorable Senate and House of Representatives of the State of Rhode Island:

This memorial of Thomas S. Ridgway, respectfully showeth and presents to your Honorable Bodies practical information in his possession, relative to the coal fields of the State of Rhode Island, laying before you facts and reasons why the coal of Rhode Island has not been more fully developed.

Dr. C. T. Jackson, many years ago, made an agricultural and geological survey of the State of Rhode Island, describing the area and boundaries of the coal formations, &c. But I come before you with information and facts obtained during three years of experience as Superintendent and General Agent of the Pocasset Coal and Iron Company, whose mine is situated in Cranston, R. I., about five miles from the city of Providence. And if I can aid in developing the mineral resources of the State of Rhode Island by means of this report, I shall feel that my labors heretofore have not been lost.

The greatest difficulty that presents itself to a practical miner in searching for a workable bed of coal in the anthracite coal field of Rhode Island is, that the best portion of the coal strata is deep seated and covered up to the depth of from 40 to 150 feet with sand, gravel, pebbles and boulders; a deposit well known amongst geologists by the name of drift or glacial formation, and also with some patches of the tertiary deposit; concealing effectually not only the outcroppings of beds of coal that may exist, but all traces of their locality. The streams of water, those efficient agents in other coal fields of the United States, exposing coal beds to view by their denuding action, have not washed their way down through the overlying sand and gravel of the State of Rhode Island, but run along on top of this sterile deposit.

The Pennsylvania, Virginia and western coal fields, are covered up with a thin deposit of alluvial, and in some places with drift. The fresh water streams there, have not only made their way down through these modern deposits, but have denuded the coal measures so deep in many places, that beds of coal are exposed to view and mined with success, high above water level. The beds of coal can be traced for miles distant, along their outcroppings by means of the depression of the land immediately over them; also by the growth of thrifty trees, and by small particles of coal and coal shale found in the sub-soil. It is not an uncommon occurrence for three or four miners to sink a test-hole in search of coal, in Pennsylvania, and in a day or two find the solid coal. But in the Rhode Island coal basin, even along and upon its upturned edges, it is a matter of chance if a dozen test-holes of moderate depth, strike into a bed of coal. It took five weeks of thought and the labor of a gang of men to open the valuable bed of coal on the Harris estate in Soccanosset Hill, Cranston, R. I.

All the discoveries of coal in the State of Rhode Island, previous to the year 1864, have been accidental—either found by sinking wells for drinking water, or by diggings made for the foundation of a house; excavations invariably made into greatly disturbed coal strata found not far beneath the surface of the land. These accidental discoveries gave rise to coal mining companies, based upon ignorance, crushed coal, and contorted wall rocks. A strong effort would be made to mine and sell the coal, and one agent would have to contend against thirty selling Pennsylvania coal. Under these disparaging circumstances, the company not being successful, would only mine for a *show*, and commence selling stock of the company instead of coal, and finally fail. Fail! there is no such thing as fail! Men may fail, but Rhode Island coal never!! How could the good people of Providence get coal after the company had failed? (for there has seldom been but one mine in operation in the State at a time,) then, by the time the same sleeping mine could be waked up by a new mining company, the people of Providence would be well stocked with Pennsylvania coal; hence the non-success and strong prejudice against the coal, and the numerous stories set afloat by ignorant persons.

It is not to be expected that a people who has been in the habit of burning wood and Pennsylvania coal for a number of years, could at first fall into the general use of Rhode Island coal, at a time when coal for domestic purposes was not prepared as it is to-day. Formerly, Rhodè Island coal was sent to market in large lumps. Here was the great error. Had the former coal operators in Rhode Island coal, studied more about combustion, they would have made a strong effort to introduce the coal for domestic purposes. For culinary purposes the coal should be the size of chestnut coal, and always consumed in a stove, as it is too compact a coal to burn in an open grate in a small furnace. The coal being broken fine, presents many surfaces for combustion, and makes a beautiful hot fire, free from slate and bony coal.

For steam purposes it should be consumed in a large body, and not over the size of ordinary stove coal, with plenty of boiler surface, and always avoiding a white heat, because it is a red ash coal.

The mining of coals in the Pennsylvania anthracite coal field is accomplished by means of slopes and pits, and some little by drifts into the beds of coal found above water level. The largest portion of coal mined comes from a great depth below water level, and is raised to the surface by steam engines. Here the beds of coal are generally regular and uniform in thickness, but not without their faults. An undulating coal bed, similar to the one mined upon Rhode Island, and variable in thickness, made up of pockets of coal, still contains the same amount of fossil fuel, as if it were uniform in thickness, needs less propping, and a less number of pillars of coal to sustain the roof of the mine, the mine being arched.

In the bituminous coal basin, near Richmond, in the State of Virginia, the beds of coal are very irregular in their direction and inclination, and vary so much in thickness that additional inclined planes, horse gins and pumps have been constructed under ground in the mine at some distance from the bottom of the elevating pit. The yield of coal from the subterranean caverns of this perplexing and irregular mining region was 100,000,000 bushels, up to the year 1850. The Midlothian Company's mining pit is nine hundred feet deep, and the Creek Company's pit, four hundred and twenty-five feet deep.

The Richmond coal field was formed at a later period of time than the Rhode Island coal formation, and reposes upon granite. It was tossed about more than the Rhode Island coal strata by igneous rocks of a later date.

The Rhode Island coal field, including the Massachusetts coal measures, is a large but shallow one, made up of a cluster of beautiful coal basins, being identical with the lower coal series of the coal measures of the anthracite coal basins of the State of Pennsylvania, the coal beds reposing in many places in an undisturbed condition, in merchantable blocks, with a good deal of regularity, and less altered by heat than along its edges.

It is well known amongst geologists, that the Rhode Island and Massachusetts coal field has been subjected to a greater degree of heat escaping from the interior of the earth, when there was a partial elevation of the land and a partial subsidence of the waters, than the Pennsylvania anthracite coal field, hence the name, "Metamorphosed Coal Field," a name derived from the burnt or baked condition of the coal and its associated strata, noticed along the margin of the field. But this burning and alteration by plutonic heat may have only changed the coal and its associated rocks near the borders of the formation, and at some points or lines of elevation in the interior. Suppose a cook should let a blackberry pie remain too long in a hot oven, yet if the edges were burnt one might make a good meal out of the interior.

The coal basin on the island of Rhode Island, extending from Bristol Neck to Newport, is a disturbed one, and yet the Portsmouth coal mine is a success, and has battled with prejudice for a number of years. It still stands the gale, and would certainly not be wrought if it did not pay. To-day, at this mine, the proprietors are consuming their coal in smelting copper ore brought from Chili, South America.

During my agency for the Pocasset Coal and Iron Company, there was 3,600 tons of coal mined and teamed from the mine in Cranston to the city of Providence, and all was sold at a less price per ton than Pennsylvania coal, in order to introduce it. It was largely consumed for steam purposes at the Hope Iron Works, G. G. Hicks' Boiler Works, American Butt Company's Works, Payton & Hawkins', and at a number of other places. The coal was mined one winter, as per contract with the miners, at seventy cents per ton, including powder, oil and fuse; add cost of running steam engine, breaker, &c., the coal was produced upon the surface, prepared for the market at $1.50 per ton. This estimate is based upon the yield of the mine at forty tons per day. All that this mine really requires is a railway connection with Providence city, then the coal can be mined and laid down in Providence at $2.50 per ton, and sold at $4.00 per ton.

But I wish to call the attention of your Honorable Bodies more particularly to the *manufacture of iron from the ore, by using Rhode Island coal.* Previous to the year 1836, there was not an iron blast furnace within the limits of the State of Pennsylvania using anthracite coal in the reduction of iron ore. The furnaces were all charcoal furnaces, making what is termed charcoal pig-iron. It is a well known fact that charcoal pig-iron sells for fifteen and twenty dollars per ton more than anthracite pig. Why? It is because charcoal does not contain sulphur, and Pennsylvania anthracite coal does. One cannot pick up a lump of the best of Pennsylvania coal and analyze it, but that it will be found to contain sulphur. Rhode Island coal does not contain sulphur in the mass at the Cranston mine; it is sometimes found in the interstices, which is immediately dissipated when the coals are thrown upon the fire. It has all the heating qualities and better fluxing ingredients in the ash for the manufacture of iron from the ore, and will also make a better quality of iron than Pennsylvania coal. The following certificates will be additional proofs to my statements:—

20 State Street, Boston, May 25, 1866.

THOS. S. RIDGWAY, Esq.,

Dear Sir:—In reply, there is not any magnesia or manganese in the *Pink Ash* coal you are mining, which can rise in the flame of that fuel. It becomes fixed in the ash, in contact with silicate and carbonate of lime. I cannot see any objection to annealing steel in its flame, as the flame is more pure than that from other coal.

Truly yours,

A. A. HAYES, *State Assayer.*

Worcester, Sept. 28, 1866.

TO THE POCASSET COAL AND IRON CO.,

Gentlemen:—We have had 15 tons of your coal from Cranston, R. I., and are using it in our Cupola Furnace in connection with Pennsylvania anthracite coal. We continue to use it and are satisfied that it improves the strength of the iron.

Yours, respectfully, &c.,

HEALD, BRITTON & FORD.

Rhode Island coal contains eighty per cent. of carbon, ten per cent. of silica, six per cent. of alumnia, and three per cent. of lime and magnesia, and during its combustion leaves a spongy, friable cinder. All coals of the anthracite quality clinker if the fires are urged up to a white heat. A white, hot fire is not necessary to produce steam, except when much work is to be done with but little boiler surface, then the consumer loses genuine heat, in some instances 60 per cent. Rhode Island coal is equal in quality to many of the red ash coals of Pennsylvania. It does not contain any more ash than many of the Pennsylvania red, grey, or white ash coals. The Cranston mine has less slate and bony coal in the mass than the majority of coals mined in Pennsylvania. It does not contain two per cent. of bony coal and slate. For three consecutive winters my family in Boston burnt coal from this mine, and in no instance did the fires die out by being choked up, as they generally do once or twice a week, when using Pennsylvania anthracite coal. Why is this? The Pennsylvania coal beds are nearly all divided into what is termed benches of coal, generally into three or four divisions, top, middle, and bottom bench; these benches or seams of coal are separated generally with a layer of slate or dirt. The miner, in breaking down the mass of coal, of course, cannot well avoid the slate commingling with the coal. Many efforts have been made in Pennsylvania to separate the two, and at some mines they have attained a fair success. But the very fact of the slate being in the mine, proves the existence of a little more slate, by a law of deposition, that slate in thin layers, as thin as paper, will be found above and below, intermixed with the pure coal, and this is called bony coal. The foregoing facts, being submitted to a candid world, are not written to decry coals that have proved themselves to be of inestimable value to the government of the United States and its people, but to set forth the intrinsic value of Rhode Island coal in its true character. Rhode Island coal is found in the bed in mass, it has not any intervening slates; if slate is found in the mined coal it comes from the bottom slate or roof of the mine; occasionally there may be a thin vein of quartz running across the coal bed, but it is so white and the coal is so black, that they can be readily separated. The quartz cannot be sold for coal.

As regards the manufacture of iron from the ore, using Rhode Island coal as fuel, I verily believe that it will make iron nearly equal to charcoal iron. An eminent chemist told me that it would take less coal than Pennsylvania anthracite to produce a ton of iron. What the State of Rhode Island really requires in the nineteenth century, is iron blast furnaces, making her own iron from the raw material in the State.

It is a mistaken idea some people have, that the coal and iron ore is generally found in Pennsylvania in juxtaposition, and that they come out of the same opening in the earth. Many iron masters in Pennsylvania transport their coal or ore 100 and 150 miles to the furnace. There are some furnaces running partially upon the ores of the coal field, but they generally rely upon outside ores brought from a distance.

If the iron ores of the coal fields of the west are so valuable, why is it that upwards of two hundred thousand tons of Lake Champlain iron ore find a market in Pennsylvania, Delaware, Ohio, Michigan, and Missouri? I will tell you why. The iron ores of the coal fields are generally found in thin bands, meagre in iron and costly to get.* But the day will come when they will be mined to an advantage; when a three-foot seam of coal is mined which was laid down providentially for future ages, long after Rhode Island is studded with iron blast furnaces.

Respectfully submitted for your consideration,

THOMAS S. RIDGWAY,

Geologist and Mining Engineer,
and Iron Blast Furnace Reviewer.

BOSTON, February, 1868.
10 Nassau St.

* Iron ores may hereafter be discovered in the Rhode Island coal field.

SUPPLEMENT.

To the Honorable Senate and House of Representatives of the State of Rhode Island:

GENTLEMEN :—I respectfully call your attention to my report upon Rhode Island coal, presented at your January Session, 1868, particularly to pages Nos. 4, 7, and 8. The certificates on page 7, relative to the quality of the coal in your State are gratuitous from the parties who have signed them. And I further remark, that I do not own an acre of mineral land in the State of Rhode Island, nor have I at present any interest in the Pocasset Coal and Iron Co.'s coal mine of Cranston, R. I.

Since that report was presented to you for consideration, I have made numerous geological examinations of the coal formation in the State of Rhode Island, and also of deposits of iron ores near at hand.

GEOLOGY OF RHODE ISLAND COAL MEASURES.

Judging from examinations already made, it appears as if nearly one-half of the coal measures in the State of Rhode Island are broken up, making what is termed amongst coal miners, " bad mining grounds." And it is upon this character of mineral land, which embraces underlying twisted and contorted strata of sandstones, slates, and crushed coal, &c., that all mining of coal heretofore within the State has been done, chiefly by beginning in some spot where the coal was accidentally discovered above water level, a point most desirable for speculative purposes, as the coal shows itself near to the surface.

The good mining grounds in the Rhode Island coal field are made up of a cluster of pools or basins of workable beds of coal, reposing below water level, in good mining condition, only waiting for capital and skill to probe it by means of boring tools and pits to win it.

These pools or basins of coal are separated, however, by the bad mining ground, composed of points and short lines of elevation of the coal and its associated strata in a much disturbed condition.

The only partially successful coal mining range of disturbed coal lands within the State, is upon the island of Newport, where the coal is found in great abundance and to exist in pockets; and yet that island, agreeable to Dr. Channing's statement, has yielded upwards of 200,000 tons of coal within twelve years past.

All that has been done, heretofore, in the shape of coal mining within the State of Rhode Island, I consider as so many test openings, proving the existence of an abundance of valuable mineral coal, suitable for steam and metallurgic purposes.

RHODE ISLAND COAL FOR STEAM PURPOSES.

The old coal mine at Portsmouth, R. I., has, in its underground construction, what is termed by miners, 2 lifts, one of 600 feet deep on the slope, and the other 1,200 feet deep. As there has been upwards of 200,000 tons of coal, and many thousand tons of rock and dirt, and an incalculable amount of water raised to the surface from these 2 lifts, by the power of steam, generated from water by consuming Rhode Island coal, I cannot see why every cotton mill in the State of Rhode Island should not use Rhode Island coal for steam purposes. Now all that is required to accomplish this, is to have more coal mines and more well-prepared Rhode Island coal in the market; and for those persons who desire to use the same for steam purposes, to construct an abundance of *boiler-absorbing surface* and *deep fire boxes*, then to exercise a moderate combustion of the coal, or use mixed coal, say one-half Pennsylvania anthracite.

Iron Ores.—There is a huge mass of iron ore at Cumberland Hill, in the State of Rhode Island. It far exceeds anything of the kind in the State of Pennsylvania, for its quality, quantity, and the facility to extract it, and nearness to a desirable mineral coal, suitable for the manufacture of iron from the ore. The intrinsic value of this ore is already known.

The State has also within its limits valuable deposits of hematite and bog iron ores.

Pig Iron.—All of the refractory iron ores of the New England States can readily be reduced to pig iron with the use of Rhode Island coal in blast furnaces. Pig iron can be made to-day in Rhode Island, from the iron ores near at hand with the use of Rhode Island mineral coal, at a cost of from $22 to $23 per ton, whereas in the State of Pennsylvania the cost of anthracite pig iron rates from $28 to $30 per ton. The largest portion of the coal or ore in Pennsylvania being transported a great distance. I know that when pig iron is produced from the New England iron ores with the use of Rhode Island coal as fuel, it will be a superior quality of iron to the Pennsylvania pig iron produced from its ores and anthracite coal.

Steel.—Pittsburg, Pennsylvania, is the principal market for American steel, which is made chiefly from hermatite iron ores, it being an inferior article to foreign steel. Steel rail road iron can be made cheaper in the State of Rhode Island (from Rhode Island pig iron), than in Pennsylvania, and may be shipped to New Orleans and South America more readily than from the interior of Pennsylvania

I do not know of any section of the country along the Atlantic coast of the United States, where good pig iron and steel can be manufactured to such an advantage as in the State of Rhode Island, where the coal field is almost in the lap of the sea, and valuable iron ore deposits forming an amphitheatre to it.

THOS S. RIDGWAY,

Geologist and Mining Engineer.

Boston, Feb. 15, 1870.
10 Nassau St.

www.ingramcontent.com/pod-product-compliance
Lightning Source LLC
LaVergne TN
LVHW020641110826
845149LV00004B/1303

* 9 7 8 1 4 1 8 1 8 9 5 9 4 *